Entrepreneurship in Chemical Engineering 2020

Topical Conference at the 2020 AIChE Annual Meeting

Online
16 – 20 November 2020

ISBN: 978-1-7138-2289-9

Printed from e-media with permission by:

Curran Associates, Inc.
57 Morehouse Lane
Red Hook, NY 12571

AIChE
120 Wall Street, FL 23
New York, NY 10005-4020

Phone: (800) 242-4363
Fax: (203) 775-5177

www.aiche.org

Additional copies of this publication are available from:

Curran Associates, Inc.
57 Morehouse Lane
Red Hook, NY 12571 USA
Phone: 845-758-0400
Fax: 845-758-2633
Email: curran@proceedings.com
Web: www.proceedings.com

TABLE OF CONTENTS

601998 What I Learned in My Industrial Postdoc: Project Management Skills for Academia

Abstract:

For many, a postdoctoral position is the next major career step in their career path after obtaining a Ph.D. A large majority of postdocs, 89%, will be in academic labs. The rest receive their training in industry, national laboratories or other government entities. While industrial postdocs are relatively rare, the experience offers a variety of benefits, including the opportunity to participate in project and grant management. Here, we specifically discuss the benefits of industrial postdocs in gaining project management experience. The hands-on training can provide strategies for assembling and managing a team, project goals, budgets, and collaborations. How those skills can be translated to successful academic grant management will be highlighted.

613042 Managing an Interdisciplinary Team of Undergraduate and Graduate Researchers

Abstract:

The success of grants is dependent on the organization of research efforts. This is especially important when the researchers have different areas of expertise, as well as different levels of experience. In this presentation I will discuss managing an interdisciplinary team of researchers from both the undergraduate and graduate level.

606309 Leading Technical Teams for Complex Energy and Sustainability Problems

Abstract:

The sustainability challenges of the 21st century are sufficiently complex to require a team approach of diverse skills and backgrounds. The development of carbon-free energy, a circular economy of recyclable materials, and long-term sustainable transportation and clean water systems will only occur with transformational approaches which combine technical, social, and economic solutions. Leading technical teams to address these problems has its own unique challenges. Academics working on these chemical technologies need to communicate and work across skill areas including research subfields of simulation and computational chemistry, process systems and analysis, materials synthesis and characterization, and experimental chemistry and catalysis. Moreover, technologies emerging from these technical teams must integrate with social expectations and economic restrictions; the teams must both share and receive feedback on the viability of their ideas for both implementation with society and commercialization across many types, sizes, and structures of companies. In this presentation, best practices and experiences are shared on maximizing the success of academic teams taking on the most important challenges of the coming decades. Specific examples include an approach to selecting problems, organizing teams, data and idea sharing, and communicating long- and short-term strategy. Finally, a discussion is provided on strategies for pivoting to new solutions and goals as insights emerge and inform the problem.

608011 Creating Independent Researchers

Abstract:

In my experience, writing a grant is typically more about scientific questions and experimental procedures than it is about the details of actuating the project in the real world. However, once the grant is funded, the challenge shifts from one of scientific inquiry to project management. This raises the challenge of achieving a stated set of scientific and/or technical goals while working to educate and empower the individuals (particularly students) who are involved in the research. Managing a research group and mentoring students is often cited as one of the most challenging aspects for new PIs. I will discuss my strategies for creating a vibrant and collaborative research group that meets the individual members where they are and empowers them to take ownership over a research project and grow as an independent scientist.

617295 Entrepreneurship & Investing in Early-Stage Chemical Companies Panel Discussion

Abstract:

This session will be a panel discussion feauring several panelists.

Judith Giordan is a co-founder of the Chemical Angels Network (www.chemicalangels.com), Managing Director of ecosVC, Inc. (www.ecosvc.com), entrepreneur and former Fortune 100 executive. Judy serves as board member, co-founder, advisor and investor in seed and early stage STEM aligned startups. Her previous executive positions include Corporate Vice President and Officer/Global Director of Research and Development at International Flavors and Fragrances, Inc.; Vice-President Worldwide Research and Development for the Pepsi-Cola Company; Vice President Research and Development/Corporate Officer, Henkel Corporation, the North American operating unit of the Henkel Group. She has also held management, program leadership and technical contributor positions at Polaroid and ALCOA.

Judy has served on Boards at the National Research Council and at the National Science Foundation where she has been a member of the Math and Physical Sciences Directorate Advisory Board, Member of the Chemistry Division Advisory Board, Member and Chair of the Waterman Award Committee, Member of External Visiting Committees, and as Program Director for the IGERT Program.

Active in Professional Societies, examples include as a member of the Board of Directors, Committee Chair and Program co-chair at the Industrial Research Institute and the American Chemical Society; a member of the advisory group of and program leader for the Conference Board technical conferences, Educational Foundation Trustee at the Commercial Development and Marketing Assoc., and speaker and program organizer at the Angel Capital Assoc. Judy's honors include ACS Fellow, ACS Garvan-Olin Medalist, ACS Henry Whelan Award for Management Leadership and Alexander von Humboldt Networking Award recipient. Judy has has served as theme co-chair for the 2020 ACS National Meeting "Moving Chemistry from Bench to Market".

Judy has a BS in Environmental Science and VoTech Agriculture from Rutgers University; PhD in Chemistry from the University of Maryland; and was an Alexander von Humboldt post doctoral fellow at the University of Frankfurt, Germany.

612375 Leading and Managing Industry-Academia Partnerships and Collaborations Panel Discussion

Abstract:

Outline:

- Discussion on benefits resulting from and concerns hindering industry-academia collaborations
- New methods and long term changes needed to lead and manage industry- academia partnerships and collaborations
- Benefits to diverse sectors including R&D, manufacturing, economics, strategic planning etc. and increasing participation across all.
- Definition of research contracts, communication channels, international collaborations, project timelines and deliverables

Invited Panelists:

Barclay Satterfield: External Innovation Leader at Eastman Chemical Company who has experience in open innovation and industry/university research partnerships, polymer property research and environmental life cycle analysis. She has been involved in a $10M 6-yr research partnership between Eastman Chemicals and NC State University.

Kenneth Rueter: President and CEO - UCOR LLC, a past Management Award recipient has over forty years of experience. He has been involved in a highly successful collaboration with the University of Tennessee Nuclear Engineering Department. This collaboration has established a Nuclear Decommissioning minor at the University of Tennessee enabling students to meet the high demand for Nuclear Legacy Cleanup around the World.

Dr. Kiarash Vakhshouri: Young professional working with Google with a solid expertise in structural characterization of semiconductor materials, thin film coating and printing, rheology, lithography and designing and operating laboratory and pilot plant chemical processes. He has several well-cited publications and four US patents to his credit and has been engaged in a number of industry-academia collaborative research projects right since graduate school.

Dr. Phil Westmoreland: Has a PhD from MIT and has been a Professor of Chemical and Biomolecular Engineering with the North Carolina State University. He has over 35 years of academic research experience and has held a variety of Directorial positions during his illustrious career. He has also served as the AIChE President in 2013. His research interests include flame chemistry, theoretical and computational chemistry and polymer decomposition kinetics and thermochemistry. He has been instrumental to a number of industry-academia-government collaborative projects.

Dr. Quinta Warren:CEO at Energy Research Consulting, PE, PMP has experience in industry, academia as well as Government. She was working with ConocoPhillips, has been a director of professional education and also worked with the US DOE. She brings in a unique entrepreneurial skill-set into synergizing these areas.

Dr. Sarma Pisupati: Professor of Energy and Mineral Engineering/Chemical Engineering and Director of Center for Critical Minerals at Penn State. He has more than 35 years experience in academia and has prior industrial experience as well. He specializes in computational fluid dynamic modeling of combustors, coal/biomass gasification, and biomass based materials for emission reduction. He has driven several successful collaborative research projects and novel initiatives with the US DOE as well as industrial partners.

Samuel Tam:(Have reached out only this week and waiting to hear back) President and CEO of CAC Solutions LLC and has been with the US DOE for 9 years where he managed total annual R&D spending was over U.S. $100 MM. He also has experience as a China Chief representative while working with the US DOE. Over his forty years of R&D, engineering, business development, pilot plant design and operation, and commercialization of technologies related to climate change in US DOE and five other major US corporations, Samuel has gained immense experience on fostering collaborations.

NOTES

NOTES

NOTES

NOTES

AUTHOR INDEX